CREATURE FEATURES: CLASSIFY ANIMALS!

IT'S A MAMMAL!

BY NATALIE HUMPHREY

Please visit our website, www.garethstevens.com. For a free color catalog of all our high-quality books, call toll free 1-800-542-2595 or fax 1-877-542-2596.

Cataloging-in-Publication Data
Names: Humphrey, Natalie.
Title: It's a mammal! / Natalie Humphrey.
Description: Buffalo, NY : Gareth Stevens Publishing, 2025. | Series: Creature features: classify animals! | Includes glossary and index.
Identifiers: ISBN 9781482466867 (pbk.) | ISBN 9781482466874 (library bound) | ISBN 9781482466881 (ebook)
Subjects: LCSH: Mammals–Juvenile literature.
Classification: LCC QL706.2 H848 2025 | DDC 599–dc23

Published in 2025 by
Gareth Stevens Publishing
2544 Clinton Street
Buffalo, NY 14224

Designer: Andrea Davison-Bartolotta
Editor: Natalie Humphrey

Photo credits: Cover Volodymyr Burdiak/Shutterstock.com; p. 5 Bildagentur Zoonar GmbH/Shutterstock.com; p. 7 (dolphin) Marti Bug Catcher/Shutterstock.com; p. 7 (horse, squirrel) Eric Isselee/Shutterstock.com; p. 7 (lemur) Natalia Paklina/Shutterstock.com; p. 7 (sloth) Passakorn Umpornmaha/Shutterstock.com; p. 9 (bottom) YuriiF/Shutterstock.com; p. 9 (top) Jim Cumming/Shutterstock.com; p. 11 Prostock-studio/Shutterstock.com; p. 13 (bottom) Wayan Sumatika/Shutterstock.com; p. 13 (top) MR.Yanukit/Shutterstock.com; p. 15 (bottom) michaelhughes91/Shutterstock.com; p. 15 (top) Kima/Shutterstock.com; p. 17 L Galbraith/Shutterstock.com; p. 19 (bottom left) Oaisu/Shutterstock.com; p. 19 (bottom right) Japan's Fireworks/Shutterstock.com; p. 19 (top) Anastasia Koro/Shutterstock.com; p. 21 (bottom left) L-N/Shutterstock.com; p. 21 (bottom right) David Turko/Shutterstock.com; p. 21 (top) Andrew Sutton/Shutterstock.com.

Printed in the United States of America

CPSIA compliance information: Batch #CS25GS: For further information contact Gareth Stevens, New York, New York at 1-800-542-2595.

CONTENTS

Boldface words appear in the glossary.

Is That a Mammal?

Mammals are a very **diverse** group of animals. There are over 400 **species** of mammals found in the United States alone. Cats, elephants, and even people are all mammals! Though these animals look different, they all have features in common.

What Makes a Mammal?

Mammals are **warm-blooded** animals. Most mammals give birth to live young. Mammals have a **backbone**. They also have **lungs** and breathe air. All mammals have hair of some kind on their bodies during their lifetime.

SLOTH
SQUIRREL
LEMUR
HORSE
DOLPHIN

Animal Fur

Hair and fur **protects** a mammal's skin from the sun. They also keep mammals warm when it's cold and cool when it's hot. Fur can help animals camouflage, or blend into their surroundings. Many mammals, such as sea otters, have **dense** coats of fur.

Human Hair

Even though people don't have dense fur like other mammals, we have hair. The hair on top of our heads is made of the same **material** as animal fur. Before human babies are born, they are also covered head to toe in hair!

Baby Mammals

Most mammals grow inside of their mothers for a few months. Some mammals carry babies for a year or more! Then, the baby mammal is born live. Many mammals only have one baby at a time. Some, such as mice, can have many more!

Monotremes

One group of mammals lays eggs! Platypuses and echidnas are monotremes, the only mammals that lay eggs. After echidnas lay their eggs, they keep them in a pouch. Platypus mothers keep theirs in a **burrow**. Then, after around 10 days, the eggs **hatch**!

ECHIDNA

PLATYPUS

Drinking Milk

Baby mammals are cared for by their parents. The babies drink milk made by their mothers. As the babies grow, they slowly start eating the same foods as their parents. They stay with their parents for a few weeks or up to a few years!

Mammal Food

Some mammals, such as cows, only eat plants. These mammals are called herbivores. Other mammals, such as tigers, eat mostly meat. These animals are called carnivores. Some mammals, called omnivores, eat both plants and meat.

Where Are They Found?

Mammals can be found around the world. Mammals such as Artic hares and polar bears live in colder places. Some mammals, such as blue whales and dolphins, are found in the ocean. Other mammals, like many **primates** and jaguars, live in warm places.

GLOSSARY

backbone: The spine, or small bones found in the middle of the back of an animal that give the body support.

burrow: A hole or tunnel made by some animals to live in or keep their young.

dense: Packed very closely together.

diverse: Differing from each other.

hatch: To break open or come out of.

lungs: The parts of an animal that take in air when it breathes.

material: Matter from which something is made.

primate: Any animal from the group that includes humans, apes, and monkeys.

protect: To keep safe.

species: A group of plants or animals that are all of the same kind.

warm-blooded: Able to keep the body at a steady temperature no matter what the outside temperature is.

FOR MORE INFORMATION

BOOKS

Brink, Tracy Vonder. *Mammals*. St. Catherines, Ontario, CA: Crabtree Publishing, 2023.

Owens, Ruth. *Let's Classify and Fold Origami Mammals*. Buffalo, NY: Enslow Publishing, 2023.

WEBSITES

Britannica Kids: Mammal
kids.britannica.com/kids/article/mammal/353414
Learn more about what features make an animal a mammal.

National Geographic Kids: Mammals
kids.nationalgeographic.com/animals/mammals
Discover more mammals found around the world.

Publisher's note to educators and parents: Our editors have carefully reviewed these websites to ensure that they are suitable for students. Many websites change frequently, however, and we cannot guarantee that a site's future contents will continue to meet our high standards of quality and educational value. Be advised that students should be closely supervised whenever they access the internet.

INDEX